Ch.4 Direct 14.96

12141107

Map Math

Learning About Latitude and Longitude Using Coordinate Systems

Orli Zuravicky

PowerMath™

The Rosen Publishing Group's
PowerKids Press™
New York

Published in 2005 by The Rosen Publishing Group, Inc.
29 East 21st Street, New York, NY 10010

Book Design: Michael Tsanis

Photo Credits: Cover, p. 5 (globe) © Digital Stock; p. 5 (Meriwether Lewis, William Clark) © Bettman/Corbis;
p. 6 © Durand Patrick/Corbis Sygma; p. 9 © Douglas Slone/Corbis.

Library of Congress Cataloging-in-Publication Data

Zuravicky, Orli.
  Map math : learning about latitude and longitude using coordinate systems / Orli Zuravicky.
       p. cm. — (PowerMath)
  Contents: What is a map? — Global direction — Latitude and longitude — Mapping global differences —
Our world.
  ISBN 1-4042-2935-3 (library binding)
  ISBN 1-4042-5133-2 (pbk.)
  6-pack ISBN 1-4042-5134-0
  1. Map reading—Juvenile literature. 2. Latitude—Juvenile literature. 3. Longitude—Juvenile literature. I.
Title. II. Series.
  GA130.Z87 2005
  912'.01'4—dc22
                              2004005590

Manufactured in the United States of America

Contents

What Is a Map?

Our world is a big place. There are 7 continents, which are divided into hundreds of countries. Most countries have cities, towns, neighborhoods, and streets. The continents are also filled with mountains, rivers, lakes, deserts, and other natural features. There are millions of places in this world that you've never visited. We can learn about these places and the geography of our world by studying maps.

There are many different kinds of maps. Some maps tell us exactly where places are, which helps us figure out how to get from one place to another. Other maps show roads and highways. Still others show the temperature, climate, and population of countries. Maps can show an area's man-made structures, like buildings, or its natural features, like canyons. A map may show just one city, town, or village, a whole country, or even all of Earth.

You can find almost anything on a map, as long as you are looking at the right kind. Before you can use maps to find your way, you have to learn how to read them. In this book, you'll learn how to use **latitude** and **longitude** to find your way around the world.

Famous explorers gathered information so that mapmakers could create a true representation of our world. In the early 1800s, Meriwether Lewis and William Clark mapped a route through land newly acquired by the United States that stretched from the Mississippi River to the Pacific Ocean.

Meriwether Lewis

William Clark

People have used maps for thousands of years. One map found in China is thought to have been made around 300 B.C. It is clearly marked with distances and shows the locations of tombs of ancient Chinese rulers. Another map dates back to the reign of the Egyptian ruler Ramses IV around 1150 B.C. It shows the routes through an area of gold mines between Egypt's Nile River and the Red Sea. A third map was discovered in the Ukraine, on the western border of Russia, and could be from 10,000 B.C.! It was engraved on a mammoth tusk and shows dwellings along a river. These maps show that people have always been interested in recording information about their world in the form of maps.

Over the years, as people learned new things about the world and discovered new forms of mathematics and science, maps became more **accurate**. Today we can use the Global Positioning System (GPS) to find our way. Twenty-four GPS **satellites** above Earth send radio signals to receivers on the ground. At any time, from any location on Earth, a GPS receiver can communicate with at least 4 GPS satellites. The GPS receivers use the satellite signals to figure out where you are, how fast you are traveling, how far you have traveled, and even a route to get you where you want to go!

Many automobiles now have GPS receivers, like the one pictured on the opposite page, to help travelers find their way.

Global Direction

Even though we now have GPS to help us find our way, it is important to know directions. The most northern place on Earth is the North Pole, and the most southern place on Earth is the South Pole. Most maps have the directions of north, south, east, and west marked on a **compass rose**. Some maps have northwest, northeast, southwest, and southeast marked as well. Let's work on our direction skills with a small area map.

Look at the picture of a city neighborhood on the opposite page. If you wanted to go to the museum after school, would you need to go north or south? First, locate the school and the museum. Trace your finger from the school to the museum. Check the compass rose to see which direction you would be moving. You would walk south from the school to reach the museum.

However, if you walked straight south, you would not find the museum. What other direction do you need to travel? You would have to walk south on Cherry Street, then west along Frederick Street to the museum. If you drew a line from the school to the museum and checked the compass rose, you would see that the museum is southwest of the school, not directly south.

Is the museum southeast or southwest of the store? Use your direction skills and the compass rose to find out.

The directions on the compass rose are often abbreviated. N means north, NE means northeast, SW means southwest, and so on.

store

school

Cherry Street

museum   Frederick Street

N
NW   NE
W   E
SW   SE
S

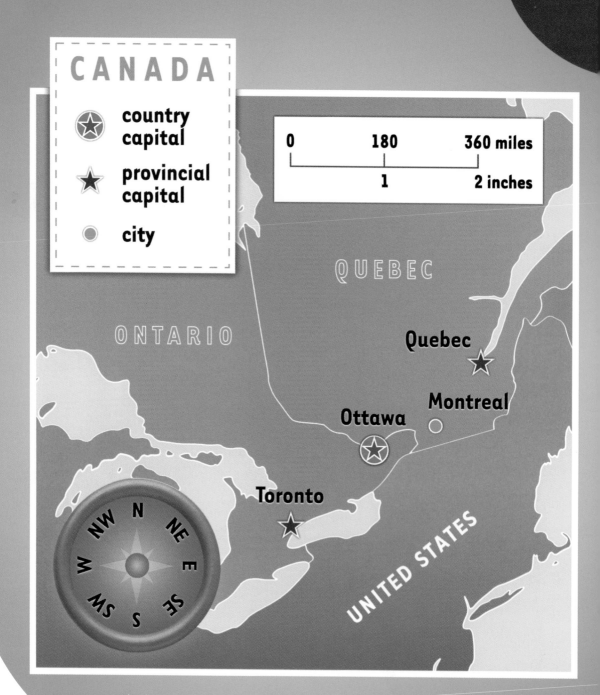

CANADA

- ⊛ country capital
- ★ provincial capital
- ◉ city

0	180	360 miles
	1	2 inches

QUEBEC

ONTARIO

Quebec ★

Montreal ◯

Ottawa ⊛

Toronto ★

UNITED STATES

Just as the United States is made up of different states,
Canada is divided into regions called provinces.

Maps can make **navigating** small areas easier. What about a map of the entire world? You have probably seen both round and flat maps of the world. Geographers and scientists decided that the only accurate way to represent the world would be on a **sphere**, which is known as a globe. However, people can't take globes with them everywhere they go. Geographers have designed flat maps that are practical for everyday use.

Of course, maps are much smaller than the places they show. To help people determine the distances between places in the real world, each map has a scale. The scale indicates the relationship between distances on the map and the corresponding actual distances. For example, a scale on a map could say that 1 inch on the map represents 1 mile on land. That way, the drawings on the map can represent the measurement of a place accurately, but in a smaller version.

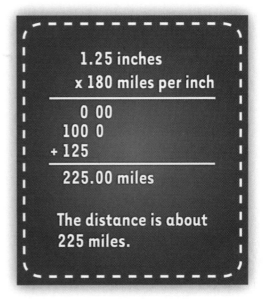

1.25 inches
x 180 miles per inch
_____
0 00
100 0
+ 125
_____
225.00 miles

The distance is about 225 miles.

Look at the scale on the map of eastern Canada on page 10. What is the approximate distance between Ottawa and Toronto on this map? First, look at the scale. One inch represents 180 miles. Next, measure the distance between the 2 cities with a ruler. The distance on the map is about 1.25 inches. To find out how many miles separate Ottawa and Toronto, multiply 1.25 by 180.

Latitude and Longitude

If you told someone that you lived on a certain street, but you didn't tell the person the city, state, or house number, they'd have some trouble finding you. It works the same way with locations on the globe. If you told someone to find a small city on a globe, it might take them a long time to locate the city among the millions of places. A system was developed by geographers to help people find places on a globe.

If you look closely at some globes and maps, you will see a lot of **vertical** and **horizontal** lines. The vertical lines are called lines of longitude, and the horizontal lines are called lines of latitude.

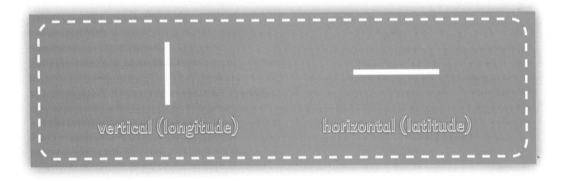

vertical (longitude)   horizontal (latitude)

These 2 sets of lines form a global grid. We use these lines to explain where on our planet specific places can be found. Numbers are given to the lines. When you combine the number for a line of latitude with the number for a line of longitude, you have coordinates, or the set of numbers used to identify a specific location. In order to understand this global grid, we need to understand how geographers decided on the system of grid coordinates.

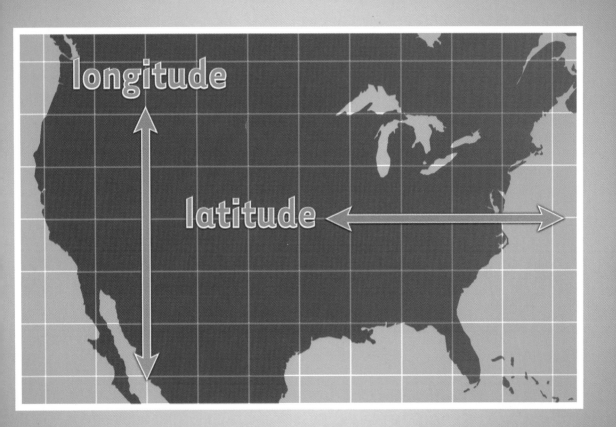

Lines of latitude are often called "parallel lines of latitude" because the distance between 2 lines is always the same. However, the distance between 2 lines of longitude becomes smaller the closer the lines are to the poles.

Earth is a sphere. It can be divided in half by drawing a line around its **equator**, making 2 equal halves. Each half is called a hemisphere (HEH-muh-sfeer), or half sphere. If you imagine a line straight through the center of Earth from top to bottom, the points at each end are called poles.

If you drew a line from the equator at the surface of Earth to Earth's center and then drew a line from Earth's center to a pole, the angle created by the two lines would measure 90°. Therefore, if we label the equator 0°, the poles could both be labeled 90°. Geographers used a similar system to create lines of latitude.

The area above Earth's equator is called the Northern Hemisphere. The area below the equator is called the Southern Hemisphere. The North Pole is 90° north, and the South Pole is 90° south. The distances between the equator and the poles are also measured in degrees and marked by lines of latitude. These lines of latitude describe positions north and south of the equator.

90°
North Pole

90°

0° equator **0°**

Earth's
center

South Pole
90°

LATITUDE

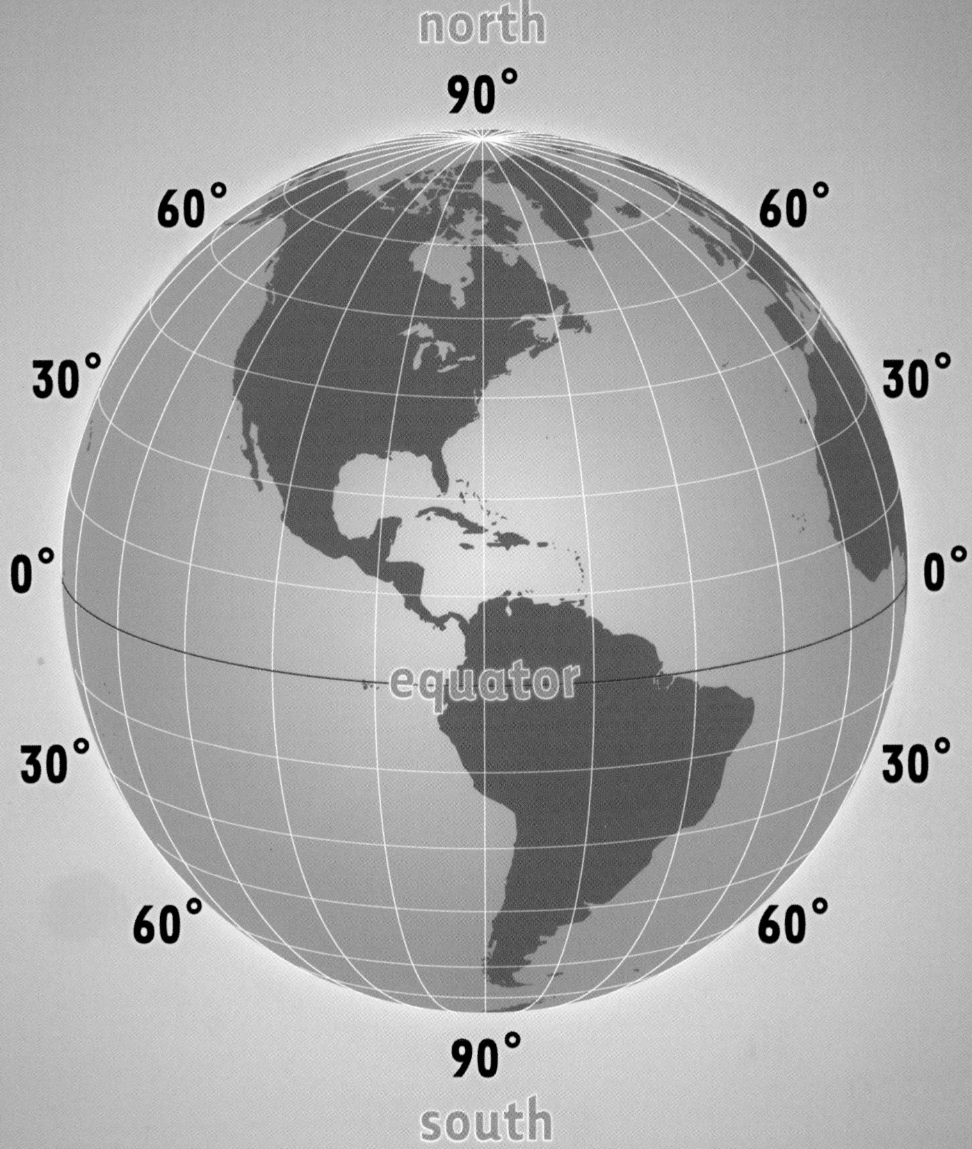

90°

60°   **60°**

30°   **30°**

0°   **0°**

equator

30°   **30°**

60°   **60°**

90°

south

If you were standing at 30° north, would you be in the Northern Hemisphere
or the Southern Hemisphere? Find the equator in the picture. Look north of
the equator. Do you see 30° north? This location is in the Northern Hemisphere.
The direction itself—30° *north*—tells us that the line is north of the equator.

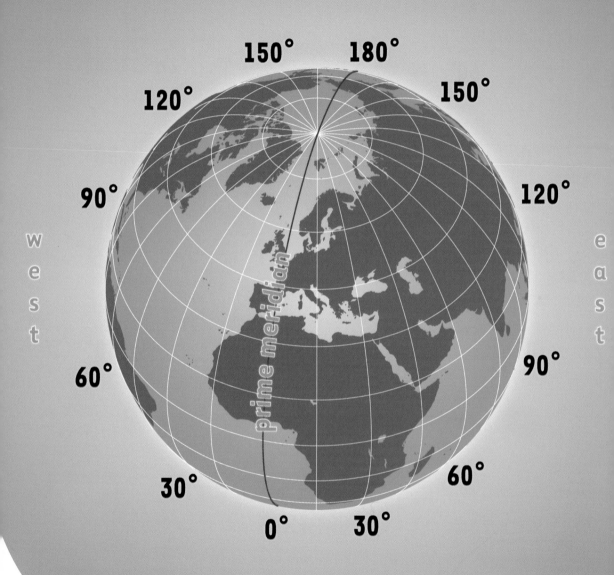

LONGITUDE

150° 180° 150°

120° 150°

90° 120°

west

east

60° 90°

prime meridian

30° 60°

0° 30°

The **circumference** of a sphere is a circle. A circle has a total of 360 degrees. Let's say you marked each degree on the circumference of a globe with a point. Then you drew a line from each point to the North and South Poles of the sphere along the surface of the sphere. When you were finished, you would have a sphere divided into 360 "slices," like the slices of an orange. This is the system that geographers used to help create lines of longitude, which are also called meridians (muh-RIH-dee-uhnz).

Lines of latitude are measured in relation to the equator and never go higher than 90°. Lines of longitude, however, begin at 0° and continue halfway around Earth to 180°. Most countries recognize the meridian passing through Greenwich (GRE-nich), England, as the 0° line of longitude, also known as the prime meridian. The prime meridian and the 180° line of longitude divide Earth into the Eastern and Western Hemispheres. Longitude measurements range from 0° to 180° east and from 0° to 180° west from the prime meridian. The location of 180° east and 180° west—which is the same line—runs through the center of the Pacific Ocean. The distance between lines of longitude at the equator is about 69 miles. The closer you go to the poles, the closer together the lines of longitude become.

If you were standing at 30° east, would you be in the Eastern or Western Hemisphere? Let's check by locating the prime meridian on the globe on the opposite page. The line of longitude 30° east would be east of the prime meridian, which places you in the Eastern Hemisphere. Again, this direction itself—30° *east*—tells us that the location is in the Eastern Hemisphere.

The lines of latitude and longitude are equally important. You need both to locate a place on the globe. Since the lines of latitude and longitude stretch across the globe, you also need to know whether the latitude is north or south and the longitude is east or west. For example, let's say you were trying to find a section of the United States. If someone told you to look on the 100° line of longitude, but did not give you a direction, you could end up looking in the middle of Asia, or in the middle of the Indian Ocean! If that person told you that it was located at 40° latitude, 100° longitude, you might still end up in the middle of an ocean. You would need the exact location—40° north, 100° west—to arrive at the right spot.

What continent are you on if you are at 20° south, 60° west? As we have learned, the direction "south" after 20° means that the line of latitude is located below, or south of, the equator. The direction "west" after 60° means that the line of longitude is located west of the prime meridian. These lines intersect in the middle of South America, which is labeled with the letter B.

What is the latitude and longitude of a place in the middle of the continent of Africa on this map? First, find Africa on the map. Since Africa is a large continent, many lines of longitude and latitude pass through it, so we'll choose the 2 lines that intersect near the middle of Africa's landmass. The line of latitude that goes through the center is between 0° and 20° north—about 10° north of the equator. The line of longitude that is nearest the middle of Africa is 20° east of the prime meridian. The address for the point in Africa that we have chosen is 10° north, 20° east.

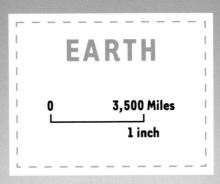

EARTH

0   3,500 Miles

1 inch

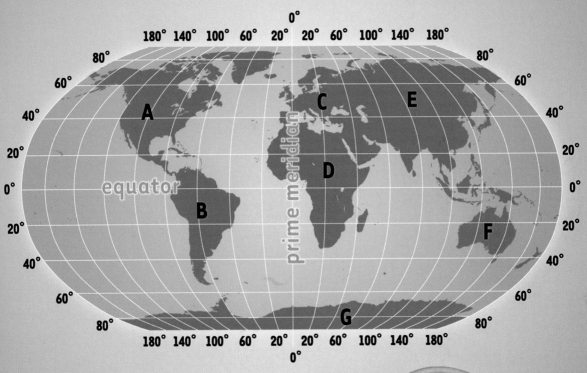

0°

180° 140° 100° 60° 20° 20° 60° 100° 140° 180°

80°   80°

60°   60°

40°   A   C   E   40°

20°   20°

0°   equator   D   0°

20°   B   F   20°

prime meridian

40°   40°

60°   60°

80°   G   80°

180° 140° 100° 60° 20° 20° 60° 100° 140° 180°

0°

KEY

A = North America   D = Africa

B = South America   E = Asia

C = Europe   F = Australia

                     G = Antarctica

MANHATTAN

0   1.9 miles

1 inch

74° 59'30" 59'00" 58'30" 58'00" 57'30" 57'00" 56'30" 56'00" 55'30" 73° 55'00"

40° 50'00"
49'30"
49'00"
48'30"
48'00"
47'30"
47'00"
46'30"
46'00"
45'30"
45'00"
44'30"
44'00"
43'30"
43'00"
40° 42'30"

HUDSON RIVER

Grant's Tomb

Harlem

Upper West Side

American Museum of Natural History

Metropolitan Museum of Art

Upper East Side

Museum of Modern Art

Grand Central Station

Empire State Building

EAST RIVER

N NE NW E W SE S SW

This map of Manhattan in New York City uses degrees, minutes, and seconds. You can understand how important seconds are in global addresses when you see how many streets in Manhattan are located between 2 minutes.

20

Maps that cover large areas, such as the whole world, may only show degrees in units of 10° or 20°. For example, they may mark 20°, 40°, 60°, but not any of the smaller degrees between. The address 40° north, 100° west will help you locate a certain section of the United States on a globe, but what do you do if you want to find the state of Connecticut? You will need a more specific address. Smaller maps, like a map of a city or state, can more easily show individual degrees because these maps focus on a smaller area. Single degrees are important for locating smaller places. For even smaller areas, different units of measurement are needed.

Since lines of longitude and latitude can be as much as 69 miles apart, each degree is divided into smaller parts. On an even smaller map, like the one on page 20, you will see lines between single degrees. Each degree is divided into 60 parts called minutes. Two minutes is written 2'. Each minute is divided into 60 parts called seconds. Five seconds is written 5". Although these words are also measurements of time, they are used to measure distance as well.

°
degree symbol

'
minute symbol

"
second symbol

Let's say you wanted to locate the global address of your friend's house in Hartford, the capital city of Connecticut. Use the map on page 23. First, locate Hartford, Connecticut. Find the line of latitude that cuts through the city. The line is more than 41°, but less than 42°. The closest minute line is between 40′ and 50′. Hartford is even a bit north of 45′, so we'll estimate the latitude is 41°46′. Now find the line of longitude. We can see Hartford is between 72° and 73°. The minute of longitude is not exactly 40′. We'll estimate the line of longitude is 72°41′.

So far, our latitude and longitude coordinates are 41°46′, 72°41′. However, we need to include directions as well. We know Connecticut is north of the equator, so the latitudinal address is 41°46′ north. We also know that this area is west of the prime meridian. Therefore, the longitudinal address is 72°41′ west. The complete address of Hartford is 41°46′ north, 72°41′ west.

Now you know the global coordinates of Hartford. But once you find the city, how will you know where your friend's house is located? The city looks small on the map, but it is much larger in reality. On a map of the city that is marked with seconds, you could be even more precise about the coordinates of your friend's house.

Almost the whole state of Connecticut lies between the 41° and 42° lines of latitude! Minutes help us pinpoint locations more accurately.

CONNECTICUT

★ capital
◉ city

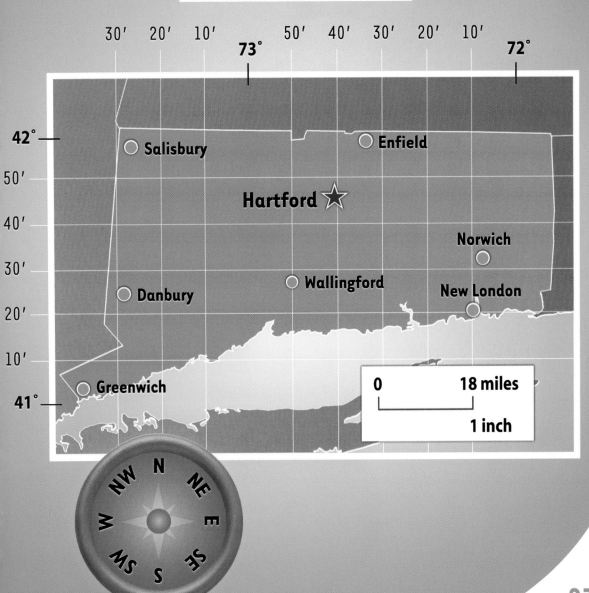

23

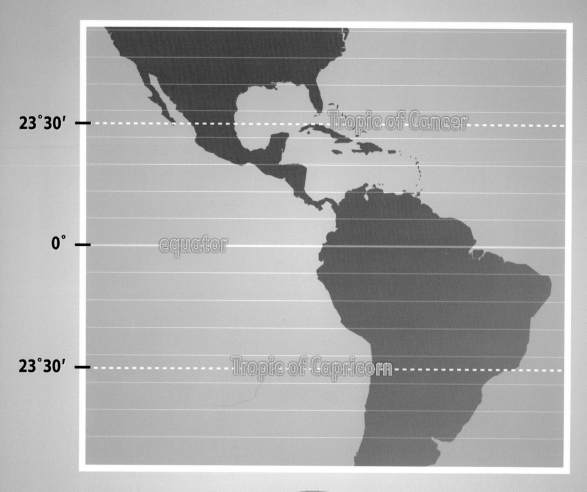

23°30' — Tropic of Cancer

0° — equator

23°30' — Tropic of Capricorn

Mapping Global Differences

Different types of maps can help explain different characteristics of each of the 7 continents, or even each country. Since Earth rotates around the sun at a certain rate, different parts of Earth get sunlight at different times and intensities. As you learned earlier, the equator divides the globe into the Northern and Southern Hemispheres. The countries north of the equator are in the Northern Hemisphere and experience winter from December to March. They normally experience summer from June to September. Countries in the Southern Hemisphere experience these seasons at opposite times of the year, with winter occurring from June to September and summer occurring from December to March.

An easy way to guess a location's climate is to compare lines of latitude, which are measured according to their distance from the equator. The sun's heat is strongest between 23°30' north (a line called the Tropic of Cancer) and 23°30' south (the Tropic of Capricorn). The sun's rays hit this area most directly, giving the area a generally warm climate during the entire year. In all other places, the sun's rays hit the surface less directly and temperatures vary during the year. Other factors, such as altitude, affect temperatures, but areas at a similar height above sea level generally can be compared according to latitude.

Which area typically experiences cooler weather: 24° north, 28° east or 60° south, 32° west? Compare the lines of latitude, the first number in each set of coordinates. The lines closer to the equator normally experience warmer weather. Therefore, 60° south, 32° west typically experiences cooler weather.

If you have family or friends who live in other parts of the world, you might have heard them say that they are a few hours "ahead of you" or "behind you." Every country falls into certain "time zones" depending on its position of longitude on the globe. Special maps show these time zones.

The lines of the time zones mostly follow the lines of longitude. The world has 24 time zones. People within the same time zone share the same time. The lines of time zones are positioned 15° apart, starting at the prime meridian. Scientists arrived at 15° because Earth takes 24 hours to complete a full rotation of 360°. They divided 360° by 24 to get the number of degrees Earth rotates per hour: 15°.

Time zones east of a specific location are "ahead," with each time zone being 1 hour later than the one west of it. The time zones that are west of a specific location are "behind," with each time zone being 1 hour earlier than the one east of it. Imagine you left New York City, New York, at 1:00 P.M. on a flight to San Francisco, California. The flight takes 6.5 hours, but clocks in the San Francisco airport tell you it is 4:30 P.M., not 7:30 P.M., when you arrive. This is because San Francisco's time zone is 3 hours "behind" New York City's time zone.

To keep the calendar in order, people created the International Date Line at 180° longitude. If you fly from west to east over the International Date Line on Tuesday, the day would then be Monday. If you travel from east to west over the International Date Line on Tuesday, the day would then be Wednesday.

180°

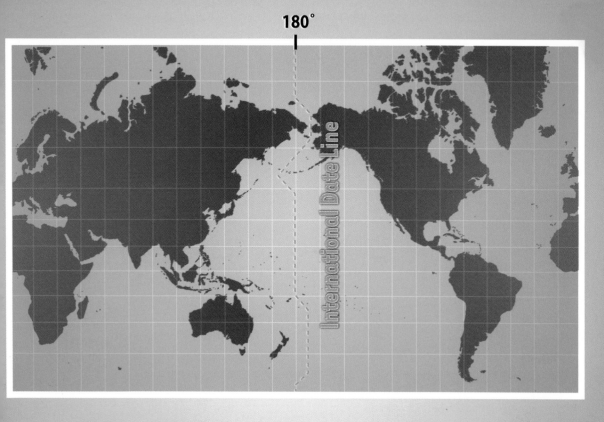

International Date Line

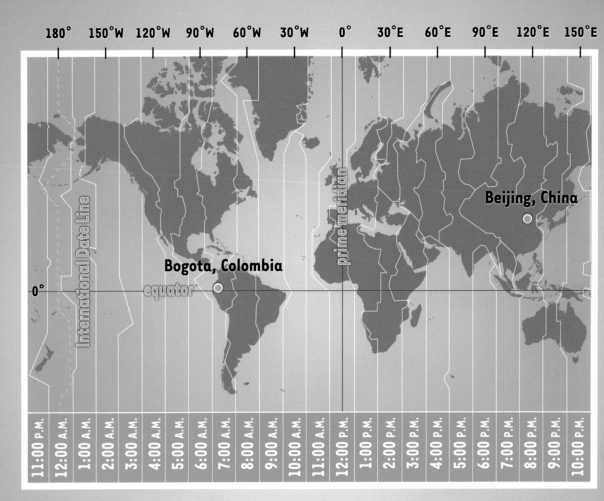

180°	150°W	120°W	90°W	60°W	30°W	0°	30°E	60°E	90°E	120°E	150°E

International Date Line

prime meridian

Beijing, China

Bogota, Colombia

equator

0°

11:00 P.M.	12:00 A.M.	1:00 A.M.	2:00 A.M.	3:00 A.M.	4:00 A.M.	5:00 A.M.	6:00 A.M.	7:00 A.M.	8:00 A.M.	9:00 A.M.	10:00 A.M.	11:00 A.M.	12:00 P.M.	1:00 P.M.	2:00 P.M.	3:00 P.M.	4:00 P.M.	5:00 P.M.	6:00 P.M.	7:00 P.M.	8:00 P.M.	9:00 P.M.	10:00 P.M.

Time zones first became necessary in the United States when trains made it possible to travel great distances in a day. Until 1883, the railroad companies in the United States used 100 time zones!

Time zones mostly follow certain lines of longitude. However, the lines curve around countries' borders because many countries adopt only 1 time zone to avoid confusion within the country. In addition, some countries within a time zone have chosen not to follow that zone's time. These factors make some time zones maps very confusing to read! The map on page 28 gives you a general idea of the different time zones of the world. We can use the map to find the time in certain areas.

When it is noon at the prime meridian, what time is it at 4° north, 74° west? Find the prime meridian. If you follow the 0° longitude line to the bottom of the map, you can see that, according to this map, it is noon at the prime meridian. Therefore, you only need to check the column in which the coordinates are located to find the time in that area.

The directions next to the degrees, north and west, tell you to look in the Northern and Western Hemispheres. The location of Bogota (boh-goh-TAH), Colombia, is marked on the map. To find the time in this part of South America when it is noon at the prime meridian, look at the column in which Bogota is located. At the bottom of the column is the time—7 A.M. Therefore, when it is noon at the prime meridian, the time at 4° north, 74° west is 7 A.M. We could say that Bogota is 5 hours "behind" the time at the prime meridian.

The capital city of China, Beijing (BAY-JING), is marked at 39° north, 116° east. What is the time in Beijing if it is noon at the prime meridian? How many hours "ahead" of the time at the prime meridian is Beijing?

PANGAEA

Asia and Europe

Panthalassa Ocean

North America

equator

Africa and Arabia

South America

India

Panthalassa Ocean

Antarctica

Australia

About 250 million years ago, scientists think that Earth's land was one big "supercontinent" that they call Pangaea (pan-JEE-uh). Due to changing ocean tides and other factors, Pangaea slowly broke apart to create the 7 continents we have today. Earth is still changing every day. Just as Ferdinand Magellan sailed around the world only 500 years ago, helping to map landforms, the scientists and explorers of today are discovering new things to help us understand the world in which we live.

Even though your home is only a small part of this big planet, it is important to keep learning about the world around us. Earth is filled with people with many lifestyles and cultures who live on different landforms. Keep practicing your latitude and longitude skills, and use them to learn about all these wonderful places. Maybe one day you'll be able to use these skills and what you've learned in your own travels around the world!

Glossary

accurate (A-kyuh-ruht) Exactly right.

circumference (suhr-KUHM-fruhns) The line around a circle.

compass rose (KUHM-puhs ROHS) A symbol placed on a chart or map to show directions.

equator (ih-KWAY-tuhr) An imaginary line around Earth's center on which every point is equally distant from the two poles. The equator divides Earth into the Northern and Southern Hemispheres.

horizontal (hohr-uh-ZAHN-tuhl) Parallel to a plane or horizon.

latitude (LA-tuh-tood) The distance north or south of the equator represented by horizontal lines, measured in degrees.

longitude (LAHN-juh-tood) The distance east or west of the prime meridian represented by vertical lines, measured in degrees.

navigate (NA-vuh-gayt) To find the course for a journey.

satellite (SA-tuh-lyt) A man-made object sent into space to orbit Earth.

sphere (SFEER) An object shaped like a ball.

vertical (VUHR-tih-kuhl) Straight up and down, forming right angles with the horizon.

Index